INTRODUCTION

Frank Schaffer's Life Science for Everyday is filled with a lot of handy, high-interest activities students can complete every day to learn about life science topics. Some of the many topics covered in this book include animals (endangered, marine, means of protection, behaviors, reptiles, insects, etc.), plants (trees, seeds, fruits, leaves, flowers, etc.), dinosaurs, human diseases, nutrition, organisms, fungi, microscopes, and many, many more.

Easy to use, many of these reproducible activities are self-contained and can be completed by students in a short amount of time at their desks. Some have been designed to help students reinforce their information-gathering skills and require quick access to such classroom-based references as almanacs, atlases, encyclopedias, and magazines. Still a few others offer students the opportunity to study a topic in a more in-depth manner. Some of the activities even build on each other and can be done in intervals over a period of several days.

Many of the activities are designed to be done individually by students. Some, however, have been created to enable students to work together in pairs, in teams, or as a class. You may want to read through each activity before assigning it to students.

Students will thoroughly enjoy learning about life science topics and concepts as they work puzzles, fill in charts, complete experiments, make observations, compare things using Venn diagrams, examine food labels, conduct interviews, take surveys, and so much more.

Regardless of whether you are trying to supplement your science curriculum, give students practice in applying scientific concepts, or encourage students to connect their own experiences to life sciences content, you will be pleased as they find the life science topics presented in this book interesting, challenging, and meaningful.

                    FS-10616 Everyday Life Science

Name_____ Date_____

EVERYDAY LIFE SCIENCE

ANIMAL ADAPTATIONS AND BEHAVIOR

Animals of all kinds have a lot of interesting ways in which they adapt to their environment and in which they behave. To learn about some of these adaptations and behaviors, match the animals and adaptations or behaviors below.

COLUMN I

1. _____ blue jay

2. _____ bobcat

3. _____ woodpecker

4. _____ crow

5. _____ tree frog

6. _____ heron

7. _____ hornet

8. _____ eagle

9. _____ turkey

10. _____ screech owl

11. _____ robin

12. _____ cicada

13. _____ opossum

14. _____ gray fox

15. _____ big brown bat

COLUMN II

A. A bird that builds a nest called an aerie

B. A large game bird that rests in trees at night

C. A blue-colored bird that squawks loudly when disturbed

D. This bird drills holes in trees in search of insects for food.

E. A fox that prefers to live in woodlands and often climbs trees

F. The males of this type of insect make loud sounds on warm summer evenings.

G. A large-eyed bird that searches for food at night

H. A marsupial that sometimes avoids danger by "playing dead"

I. An amphibian that has stick pads on its feet to help it climb trees

J. An all-black bird that caws

K. A wild cat that catches its prey by pouncing on small animals

L. A furry mammal that has a wingspan of about 12 inches

M. A long-legged wading bird that nests in flocks

N. A stinging insect that builds a paper nest

O. A red-breasted bird whose arrival signals the beginning of spring

 FS-10616 Everyday Life Science

everyday life science

ANIMAL ADAPTATIONS

Animals have adapted to climatic changes, changes in food supply, and changing landforms in order to survive. Choose three animals listed below that have had to adapt their lifestyles in order to survive. Research them and write about them below.

POLAR BEAR	WALKING STICK	CAMEL	GIANT PANDA	OWL
HUMMINGBIRD	PELICAN	SLOTH	CHIMPANZEE	FLYING SQUIRREL
EARTHWORM	KOALA	GIRAFFE	HUMAN	DEER

1. animal_____

    special adaptations:_____

_____

_____

_____

2. animal_____

    special adaptations:_____

_____

_____

_____

3. animal_____

    special adaptations:_____

_____

_____

_____

EVERYDAY **life science**

Animals' Warning Colors

Whereas many animals are protected by their camouflage and blended colors, other animals call attention to themselves with their bright colors.

List animals whose color combinations of black, white, red, orange, yellow, green, or purple protect them from predators and humans.

_____

_____

_____

_____

Look through old magazines for pictures of animals whose bright colors offer them protection. Cut out and glue the pictures in the squares below.

Name_____ Date_____

ANIMAL TRACKS

It is often possible to identify an animal by the tracks it leaves. Identify the animals' names in the spaces below by "tracking" the letters in the correct order. The first letter of each animal's name is circled. No tracks will cross in each space.

1. A Ⓞ N N R A T U G	**2.** A N Z P M H I Ⓒ E E	**3.** N T A Ⓜ E E A
4. Ⓒ L H E O M N A E	**5.** R T O O I S Ⓣ E E	**6.** O Ⓒ M O T N U T O T H
7. R C I N Ⓟ E U P O	**8.** Ⓡ H O S R O E C I N	**9.** D I P E P R A Ⓢ N
10. Ⓑ B I L D R A K D C	**11.** G N Ⓚ A R A O O	**12.** O U N D Ⓖ H O G R
13. Ⓗ G U N M B I R D I M	**14.** Ⓗ O O I A P M S U T P P	**15.** A L M N Ⓢ A D E R A

4

EVERYDAY LIFE SCIENCE

EVERYDAY LIFE SCIENCE

ANIMAL SPECIES

Carolus Linnaeus, a Swedish botanist, developed a classification system that is used today in the identification of plants and animals. Latin words are used for the last two parts of the system and represent the genus and species of the organism.

Examine the Latin terms below and match them with the animals' common names.

A. gray wolf	**G.** robin	**L.** lynx	**Q.** coyote
B. blue crab	**H.** lion	**M.** gorilla	**R.** ruffed lemur
C. bobcat	**I.** orangutan	**N.** human	**S.** house cat
D. polar bear	**J.** dog	**O.** tiger	**T.** blue jay
E. leopard	**K.** chimpanzee	**P.** giraffe	**U.** Asian elephant
F. squirrel monkey			

1. _____ Lynx rufus

2. _____ Felis domesticus

3. _____ Canis lupus

4. _____ Turdus migratorius

5. _____ Callinectes sapidus

6. _____ Giraffa camelopardalis

7. _____ Panthera pardus

8. _____ Canis familiaris

9. _____ Lynx canadensis

10. _____ Cyanocitta cristata

11. _____ Ursus maritimus

12. _____ Homo sapiens

13. _____ Lemur variegatus

14. _____ Saimiri sciureus

15. _____ Pongo pygmaeus

16. _____ Pan troglodytes

17. _____ Gorilla gorilla

18. _____ Canis latrans

19. _____ Elephas maximus

20. _____ Panthera leo

21. _____ Panthera tigris

Which animals were the easiest to identify? _____

_____

_____

• • • • • • • • •
E V E R Y D A Y

DESCRIPTIVE ANIMALS

The world is filled with a lot of awesome animals.
Read the clues below to identify the name of each animal.

1. I live in a high place. _____ gorilla

2. I have a head-covering like a king. _____ crane

3. I react to funny jokes. _____ hyena

4. I have lost my hair. _____ eagle

5. I am very large in size. _____ panda

6. My head could be a tool. _____ shark

7. My head is the color of a penny. _____ snake

8. My tail is like a baby's toy. _____ snake

9. I live in open spaces out west. _____ dog

10. My color is the same as that on a stop sign. _____ fox

11. I do not quack with my bill. _____ platypus

12. I soar like a glider. _____ squirrel

13. I have a royal name. _____ butterfly

14. My shape looks like something that is seen in the night sky. _____ fish

15. I taste good with peanut butter and bread. _____ fish

16. I appear in the sky as an arc after rain. _____ trout

17. I live in tropical waters. _____ horse

18. I live in the cold Arctic. _____ bear

19. I live in tiny grains of rocks in the sea. _____ dollar

20. I cover a lawn. _____ hopper

21. I have a hot flame. _____ fly

22. I could be a member of the Armed Services. _____ ant

23. I live on our planet. _____ worm

24. I like to live alone. _____ crab

EVERYDAY LIFE SCIENCE

ENDANGERED ANIMALS

An endangered animal is one whose population is so low that it runs the risk of becoming extinct.

Unscramble the names of the endangered animals below. The circled letters will then spell out the name of an endangered animal from Borneo.

L A O K A

G U A J A R

A M U P

A N T E A M E

T I G A N D A P A N

U C I V A Ñ

H E T H A C E

I T C A O

G H P O O R N R N

Print the name of the secret endangered animal below.

_____

From a reference book, find some interesting facts about this endangered animal.

_____

_____

_____

_____

_____

ENDANGERED ANIMAL CLUES

In a group, divide the names of the endangered animals below. Write the name of each one you choose on a 5″ × 7″ index card. Research your chosen animals and list 20 facts about them on the back of each card.

Attach your fact cards to a bulletin board along with all of your friends' cards and with a list of the animals. Take turns reading the facts and matching them to the correct animals.

AYE-AYE	BALD EAGLE	BARRED OWL
BENGAL TIGER	BLACK LEMUR	BROWN HYENA
GIANT ANTEATER	GIANT PANDA	GRIZZLY BEAR
INDIAN COBRA	KIT FOX	KOALA
KODIAK BEAR	LOWLAND GORILLA	MANED WOLF
MOUNTAIN GORILLA	OCELOT	OKAPI
ORANGUTAN	POLAR BEAR	SEA OTTER
SIBERIAN TIGER	SLOTH BEAR	SNOW LEOPARD
SPIDER MONKEY	SPOTTED BAT	TULE ELK
VICUÑA	WHITE RHINOCEROS	WHOOPING CRANE

The Most Popular Cats

Cats are favorite pets all over the world. Conduct a survey in your school to determine the most popular cats students have as pets in their homes. You may want to assemble into teams to conduct the survey in different classrooms. Write the totals in the spaces.

_____ Black Persian	_____ Blue-Cream Persian
_____ Red Tabby Persian	_____ Red Persian
_____ Cream Persian	_____ Smoke Persian
_____ Tortoiseshell Persian	_____ Blue-eyed White Persian
_____ Copper-eyed White Persian	_____ Blue Persian
_____ Brown Tabby Persian	_____ Silver Tabby Persian
_____ Shaded Persian	_____ Silver
_____ Shell Cameo	_____ Shaded Cameo
_____ Himalayan	_____ American Shorthair, black
_____ American Shorthair, white	_____ American Shorthair, blue
_____ American Shorthair, cream	_____ American Shorthair, calico
_____ American Shorthair Tabby	_____ Mackerel-striped Tabby
_____ Russian Blue Shorthair	_____ Manx
_____ Abyssinian	_____ Seal Point Siamese
_____ Chocolate Point Siamese	_____ Blue Point Siamese
_____ Lilac Point Siamese	_____ French Birman
_____ Angora	_____ Burmese
_____ Havana Brown	_____ Rex
_____ Sphinx	_____ Maine Coon

MARINE LIFE

The oceans are teeming with living things. Complete the word grid below to learn the names of some marine life.

CONCH	EEL	LOBSTER	OCTOPUS	SHRIMP	STARFISH
CORAL	HERMIT CRAB	MANATEE	SCALLOP	SPONGE	TRITON
DOLPHIN	LIMPET	MUSSEL	SEAL	SQUID	WHALE

FS-10616 Everyday Life Science

EVERYDAY LIFE SCIENCE

EVERYDAY LIFE SCIENCE

MOLLUSKS

Mollusks are animals with soft, boneless bodies. Most of them have shells. Three of the most common classes of mollusks are bivalves, gastropods, and cephalopods.

Choose one of the mollusks below to research. Data collected on each example could include the following: size, habitat, description, uses of, and unusual characteristics or behaviors. Write the data in the chart below.

BIVALVES:
oyster, clam, scallop, mussel, cockle, ark, angel wing, jewel box, jingle, ox heart

GASTROPODS:
land snail, abalone, conch, slug, sea slug, limpet, sea snail, moon snail, cone shell, murex, olive, cowrie, whelk, bonnet, periwinkle

CEPHALOPODS:
squid, octopus, nautilus, cuttlefish

Mollusk I will research (Circle one):

BIVALVE   GASTROPOD   CEPHALOPOD

Name of mollusk: _____

Type of Data:

Size:	
Habitat:	
Description:	
Uses of:	
Unusual Characteristics or Behaviors:	

FABULOUS SEASHELLS

Seashells are a lot of fun to collect and examine. What once lived in these shells? Learn the names of some shells by completing the word grid below.

CLAM	COWRIE	MOON	PERIWINKLE	TRITON
CONCH	JINGLE	MUREX	SCALLOP	TULIP
CONE	LIMPET	NERITE	SCOTCH BONNET	WENTLETRAP
COQUINA	MARGIN	OLIVE	SLIPPER	WHELK

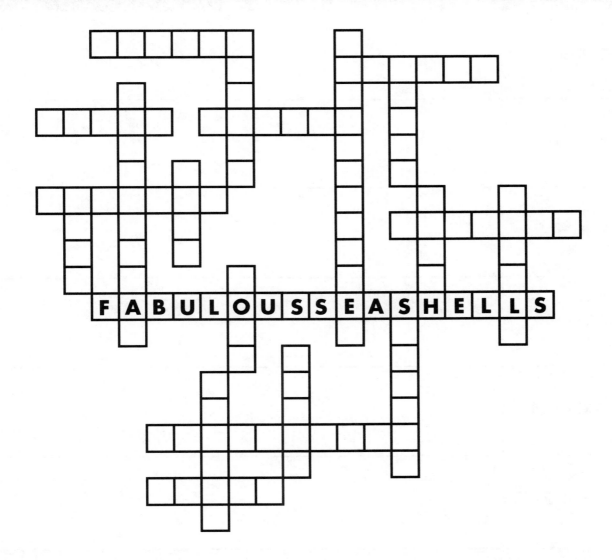

FABULOUSSEASHELLS

EVERYDAY LIFE SCIENCE

ECHINODERMS

Echinoderms are marine animals. The term *echinoderm* is derived from two Greek words, *echinos* meaning "spiny" and *derma* meaning "skin."

In a group, research the echinoderms listed below. Look for these special terms as you research your animals.

| RADIAL SYMMETRY | TUBE FEET | ARMS | WATER CANALS |
| STOMACH | SUCTION CUPS | RAYS | EYESPOT |

ECHINODERM	CHARACTERISTICS, BEHAVIORS, HABITATS, USES
STARFISH	
SAND DOLLAR	
SEA URCHIN	
SEA LILY	
SEA CUCUMBER	
BRITTLE STAR	

everyday life science

SOME SHARKS TO STUDY

Sharks live in oceans throughout the world. They are one of the most feared sea animals. Choose one of the sharks below to research. Fill in some interesting facts about the shark.

ANGEL SHARK	BASKING SHARK	BLUE SHARK
BULL SHARK	CAT SHARK	COOKIE-CUTTER SHARK
DWARF SHARK	FRILL SHARK	GREAT WHITE SHARK
HAMMERHEAD SHARK	HORN SHARK	LEMON SHARK
LEOPARD SHARK	MAKO SHARK	TIGER SHARK

Shark Name:

Size:

Description:

Habitat:

Behaviors:

Other:

EVERYDAY **life science**

Comparing Whales and Dolphins

A Venn diagram is a great tool that can be used to compare things. Use the one below to compare whales and dolphins. Write characteristics of each animal that are unique to it in the part of the circles that do not overlap. Characteristics the animals share go in the overlapping area.

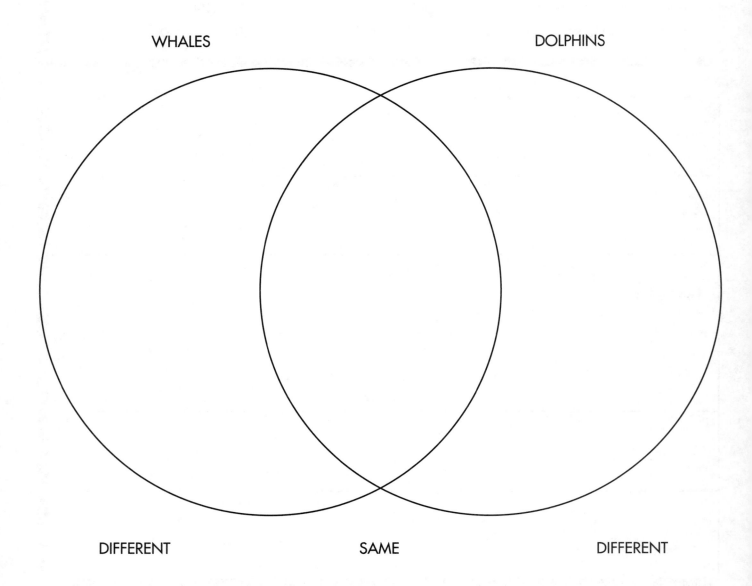

WHALES

DOLPHINS

DIFFERENT   SAME   DIFFERENT

SEAFOOD RECIPES

There are many "delicious" animals living in the sea. Choose one of the animals of the sea used as food listed below. Search different cookbooks for recipes which contain your animal. Select one recipe and print it in the space below. These recipes can be used for a class collection cookbook.

| TUNA | SALMON | OYSTER | SHRIMP | CLAM |
| CRAB | LOBSTER | SCALLOP | SQUID | TURTLE |

EVERYDAY LIFE SCIENCE

A SPECIAL GROUP OF ANIMALS

Some animals have backbones, and some do not. Unscramble the letters to spell the names of some animals that have backbones. Print the answers in the spaces to the right. The circled letters will then identify what animals that have backbones are called. Print those letters at the bottom of the page.

1. A V E R N   ___ ___ Ⓞ ___ ___

2. P E N A T E L H   ___ ___ Ⓞ ___ ___ ___ ___ ___

3. E P R A L O D   ___ ___ ___ ___ Ⓞ ___ ___

4. R A G A N O N U T   ___ ___ ___ ___ ___ ___ Ⓞ ___ ___

5. L E R U V U T   ___ ___ ___ ___ ___ ___ Ⓞ ___

6. A C O B B T   ___ ___ Ⓞ ___ ___ ___ ___

7. N I R A P R E T   ___ ___ ___ Ⓞ ___ ___ ___ ___

8. R U O A C I B   Ⓞ ___ ___ ___ ___ ___ ___

9. H E H A C E T   ___ ___ ___ Ⓞ ___ ___ ___

10. N O K Y M E   ___ ___ ___ Ⓞ ___ ___

11. R I H S O T C   Ⓞ ___ ___ ___ ___ ___ ___

What type of animal are you and the rest of the animals listed above?

___ ___ ___ ___ ___ ___ ___ ___ ___ ___

What is the meaning of this word? _____

_____

CONCEPT MAP OF VERTEBRATES

Vertebrates are animals that have backbones. Learn about them by completing a concept map of vertebrates on a sheet of posterboard. To complete a concept map, write one of the subgroups from the box in an empty oval. Then branch off each subgroup and write group names. (An example has been done for you.) Include pictures and labels in your map.

| BONY FISH | CARTILAGE FISH | JAWLESS FISH | AMPHIBIANS |
| REPTILES | | BIRDS | MAMMALS |

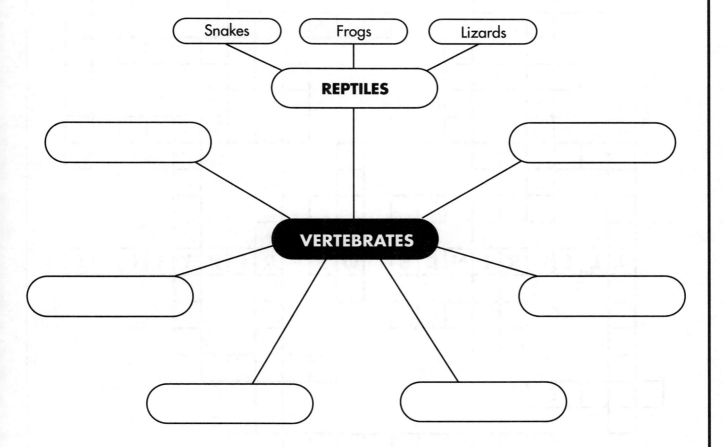

EVERYDAY LIFE SCIENCE

ALL KINDS OF REPTILES

A reptile is an animal that has dry, scaly skin and breathes using lungs. It is a vertebrate. Complete the word grid by filling in the squares with the names of the reptiles below.

ADDER	CAIMAN	GAVIAL	LEOPARD FROG	PYTHON
ALLIGATOR	COBRA	GECKO	LIZARD	SPRING PEEPER
ANOLE	CORN SNAKE	GREEN TOAD	MAMBA	TUATARA
BOA	CROCODILE	IGUANA	NEWT	TURTLE
				VIPER

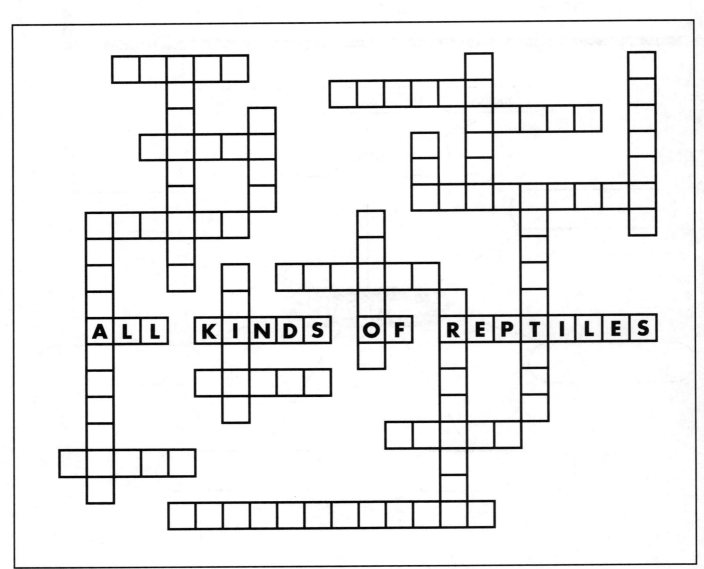

everyday life science

CONCEPT MAP OF INVERTEBRATES

Invertebrates are animals that have no backbones. Learn about them by completing a concept map of invertebrates on a sheet of posterboard. To complete the concept map, write one of the subgroups from the box in an empty oval. Then branch off of each subgroup and write group names. (An example has been done for you.) Include pictures and labels in your map.

| SPONGES | CNIDARIANS | FLATWORMS | ROUNDWORMS |
| MOLLUSKS | SEGMENTED WORMS | ARTHROPODS | ECHINODERMS |

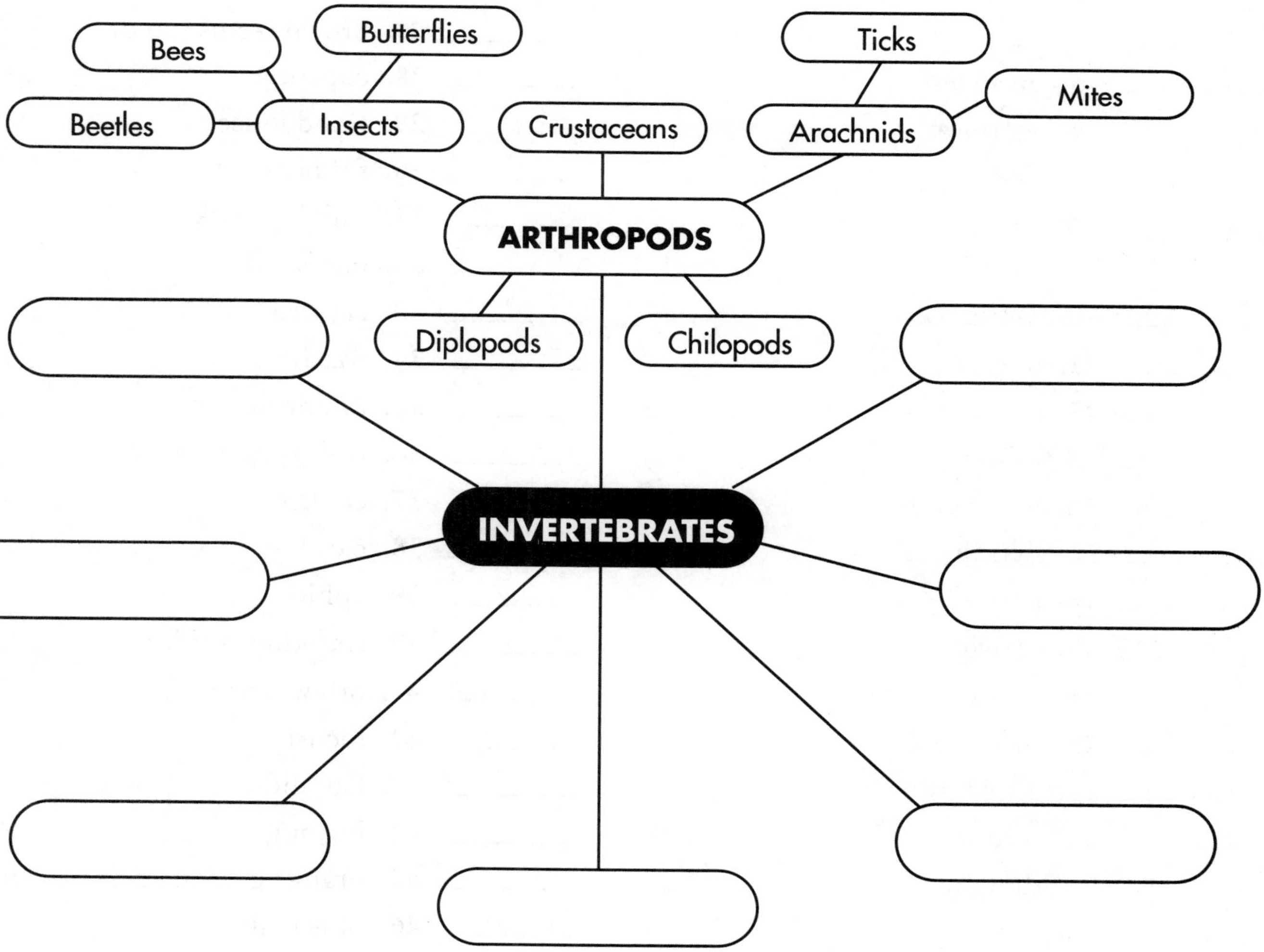

Bees
Butterflies
Ticks
Beetles
Insects
Mites
Crustaceans
Arachnids

ARTHROPODS

Diplopods
Chilopods

INVERTEBRATES

EVERYDAY **life science**

Can You Identify These Arthropods?

Arthropods are animals with segmented appendages. They are invertebrates as they have no backbones. Three of the classes of arthropods are arachnids, insects, and crustaceans. Identify the animals below by placing **A** in front of the arachnids, **I** in front of the insects, and **C** in front of the crustaceans.

_____ **1.** grasshopper

_____ **2.** mite

_____ **3.** brine shrimp

_____ **4.** wasp

_____ **5.** bumblebee

_____ **6.** scorpion

_____ **7.** tick

_____ **8.** spider crab

_____ **9.** crayfish

_____ **10.** butterfly

_____ **11.** hornet

_____ **12.** cockroach

_____ **13.** lobster

_____ **14.** water bug

_____ **15.** hermit crab

_____ **16.** barnacle

_____ **17.** earwig

_____ **18.** beetle

_____ **19.** mole cricket

_____ **20.** stinkbug

_____ **21.** termite

_____ **22.** dragonfly

_____ **23.** ant

_____ **24.** silver fish

_____ **25.** louse

_____ **26.** black widow spider

_____ **27.** brown recluse spider

_____ **28.** copepod

_____ **29.** wood louse

_____ **30.** fiddler crab

_____ **31.** walking stick

_____ **32.** damsel fly

_____ **33.** cat flea

_____ **34.** roach

_____ **35.** tarantula

_____ **36.** wolf spider

_____ **37.** cricket

_____ **38.** cicada

_____ **39.** aphid

_____ **40.** trap-door spider

_____ **41.** orb weaver

_____ **42.** locust

_____ **43.** katydid

_____ **44.** bedbug

_____ **45.** orange garden spider

_____ **46.** blue crab

THE WORLD OF ARTHROPODS

The largest percentage of animals in the world are arthropods. Arthropods are animals with exoskeletons and jointed appendages. They live in all parts of the world and in every type of habitat.

Look at the list of arthropods below. Print the names of insects in the square, the names of arachnids in the triangle, and the names of crustaceans in the circle.

TARANTULA	BEE	MITE	CRAB
LOBSTER	BUTTERFLY	SCORPION	WASP
BEETLE	SHRIMP	HORNET	FLY
GRASSHOPPER	CRICKET	CRAYFISH	TICK
GARDEN SPIDER	BROWN RECLUSE	BLACK WIDOW	CICADA
BARNACLE	LOUSE	WATER FLEA	APHID
TERMITE	ANT	WOOD LOUSE	FLEA
MOTH	FIREFLY	GNAT	MAYFLY

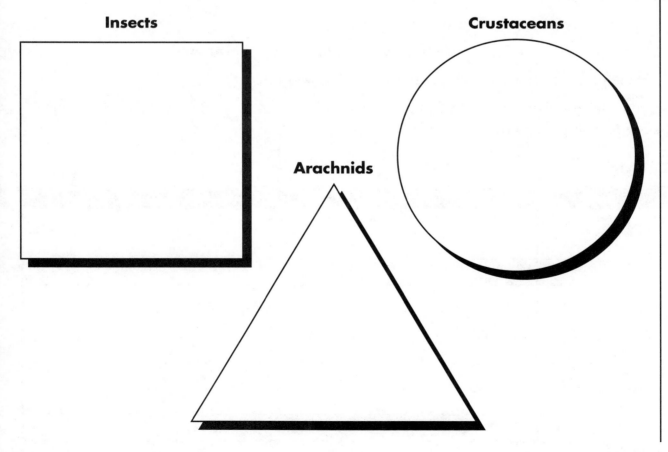

Insects

Crustaceans

Arachnids

EVERYDAY LIFE SCIENCE

Name_____   Date_____

MEALWORM BEHAVIOR

Beetles are a large and important group of insects. The name *beetle* means *biter* and refers to the strong parts of the beetle's mouth.

In this activity, you will observe the behavior and life cycle of the mealworm beetle.

> **Materials Needed:** small plastic or cardboard shoebox, bran flakes, paper towels, mealworms, water, sliced potatoes

DIRECTIONS:

1. Obtain a small carton of mealworms from a bait shop or pet store. The mealworms will be in the larval stage of the mealworm beetle.

2. Pour bran flakes into the shoebox to a depth of about one inch.

3. Place some sliced potatoes on the bran to provide food and moisture for the mealworms.

4. Add the mealworms to the box.

5. Place several wet paper towels on the mealworms, potatoes, and bran.

6. Set the box aside in a cool, shaded place.

7. You can observe the mealworms daily by carefully lifting the paper towels.

8. Add more potatoes as needed, and keep the paper towels moist.

9. Record your observations weekly. Note the changes of the larvae into pupae and adult beetles. The beetles will lay eggs to begin a new life cycle.

OBSERVATIONS:

DATE	OBSERVED BEHAVIOR AND CHANGES IN APPEARANCE

• • • • • • • • • • •
E V E R Y D A Y

TRACKING A MEALWORM'S MOVEMENTS

A mealworm is a kind of beetle that often infests such grains as flour. In this activity, you will observe and track a mealworm's movement.

DIRECTIONS:

1. Draw a dot in the center of a large sheet of white cardboard or posterboard. This will be the starting point for the mealworm's movements.

2. Place the assorted objects on the cardboard.

Materials Needed: large sheet of white cardboard or posterboard, colored pens, assorted objects (for example, a penny, a pencil, a pebble, a twig, etc.), mealworm

3. Place a mealworm on a chosen starting point.

4. Observe the mealworm's movements. Use a pen to trace the path of the mealworm as it moves across the cardboard.

5. Note the action of the mealworm as it touches the object. Does it crawl over the object, or does it turn to go around the object?

6. Repeat the trials by placing the mealworm at the starting point again. Use a different-colored pen to trace the path in each trial.

OBSERVATIONS:

TRIAL	TYPE OF PATH TAKEN, BEHAVIOR AT POINT OF REACHING OBSTACLE
1	
2	
3	
4	
5	

EVERYDAY LIFE SCIENCE

EARTHWORMS' SOIL CONDITIONING

Earthworms love to dig in soil. In this activity, you will observe earthworms and their role in mixing soil.

Materials Needed:

one-gallon glass jar, potting soil, builder's sand, chopped potatoes, crushed leaves, earthworms

DIRECTIONS:

1. Obtain earthworms from a bait shop or pet store.

2. Alternate layers of builder's sand, potting soil, and crushed leaves in the glass jar.

3. Place several earthworms in the jar. Add some chopped potatoes.

4. Place the jar in a dark cool place or cover it with a cloth.

5. Remove the cloth and observe the conditions in the jar each day.

DATE	OBSERVATIONS

How has the soil been changed by the earthworms' movements? _____

_____

Of what value are earthworms to the soil in a garden or flower bed? _____

_____

everyday life science

SNAIL OBSERVATIONS

Snails have a soft body which is usually covered with a coiled shell. Land snails live under logs and stones, on the edges of ponds and rivers, and in woods. In this activity, you will create a habitat for land snails and observe their behavior.

Materials Needed: clear gallon jar, chlorine-free water, land snails, aquarium plants, aquarium sand or gravel, magnifying lens

DIRECTIONS:

1. Obtain some land snails, aquarium plants, and sand or gravel from a pet store.
2. Fill a clear gallon jar almost full with chlorine-free water.
3. Add the plants and snails to the jar.
4. Complete the observation chart below for two weeks.

DAY	OBSERVATIONS
1	
2	
3	
4	
5	
6	
7	
8	
9	
10	
11	
12	
13	
14	

Animal-Flower Connections

Many flowers' names contain the names of animals. Unscramble the letters below to identify the animals' names that are also part of the flowers' names.

1. L A N D C A R I   ___ ___ ___ ___ ___ ___ ___   flower

2. S C T L O   ___ ___ ___ ___ ___   foot

3. G D O   ___ ___ ___   bane

4. K K U N S   ___ ___ ___ ___ ___   cabbage

5. S K A E N   ___ ___ ___ ___ ___   root

6. L E A F   ___ ___ ___ ___   bane

7. D O A T   ___ ___ ___ ___   flax

8. K O N M E Y   ___ ___ ___ ___ ___ ___   flower

9. R T G I E   ___ ___ ___ ___ ___   lily

10. N I O L   dande ___ ___ ___ ___

11. O F X   ___ ___ ___   glove

12. L E A S   golden ___ ___ ___ ___

13. B A L M   ___ ___ ___ ___   's-quarters

14. B R I D   ___ ___ ___ ___   's-foot trefoil

15. W O C   ___ ___ ___   slip

    FS-10616 *Everyday Life Science*

ANIMALS AND PLANTS LIVING IN BIOMES

Plants and animals of the world live in many different kinds of biomes. A biome is a community of plants and animals that lives in a large geographical area that has a similar climate. Identify the biome of each of the animals and plants listed below. Use the following letters for the identifications:

D desert	**G** grassland	**R** tropical rain forest
T taiga	**TU** tundra	**F** temperate forest

1. _____ low shrubs

2. _____ aardvark

3. _____ cactus

4. _____ sagebrush

5. _____ moose

6. _____ grasses

7. _____ oak

8. _____ arctic fox

9. _____ rattlesnake

10. _____ lichens

11. _____ parrot

12. _____ prairie dog

13. _____ snowshoe hare

14. _____ maple

15. _____ kapok tree

16. _____ gibbon

17. _____ caribou

18. _____ conifers

19. _____ pronghorn

20. _____ roadrunner

21. _____ tree ferns

22. _____ elm

23. _____ lemming

24. _____ kangaroo rat

25. _____ brown bear

26. _____ Gila monster

27. _____ sloth

28. _____ jaguar

29. _____ polar bear

30. _____ salamander

31. _____ zebra

32. _____ scorpion

33. _____ monkey

34. _____ African elephant

35. _____ dingo

36. _____ jack rabbit

37. _____ skunk

38. _____ ostrich

39. _____ Indian elephant

40. _____ mahogany

41. _____ squirrel

42. _____ reindeer

43. _____ spruce tree

44. _____ giraffe

45. _____ raccoon

46. _____ fir tree

Name _____   Date _____

PRODUCERS AND CONSUMERS

Producers are organisms, such as green plants, that use light energy to make food from carbon dioxide and water in the presence of chlorophyll. Consumers are organisms, such as animals, that are unable to make their own food and must obtain food from other organisms.

Distinguish between producers and consumers by placing each of the organisms below in its correct column in the chart.

TOMATO	VULTURE	POTATO	SUNFISH	SNAIL
BEET	CABBAGE	APRICOT	PEACH TREE	OAK TREE
CARROT	BEE	PECAN	TURKEY	ELEPHANT
FOX	WOLF	SHEEP	DEER	ELK
SPIDER	PIG	APPLE	BAMBOO	MOSS
IVY	ASPARAGUS	CELERY	SEAWEED	WHEAT
PORPOISE	MANATEE	BOBCAT	CACTUS	RICE
COW	SPINACH	GRASS	MOLE	HUMMINGBIRD
SQUIRREL	EARTHWORM	EAGLE	WASP	CLOVER

PRODUCER	CONSUMER

On another sheet of paper, use some of these examples of producers and consumers to construct a food chain.

• • • • • • • • •
E V E R Y D A Y

TREES IN USE

Trees are very important to mankind. They give people many things and are used for many things. Use old magazines, newspapers, catalogs, or sales supplements to find pictures which represent the many uses of trees and products made from them. Cut and attach the pictures in the spaces below. On the back of this page, include any other uses of trees and attach additional pictures. Compare your collection with those of other students.

FISHING	SKIING	PAPER	CHARCOAL
MATCHES	DOORS	SHOES	CORK
FRUITS	TOYS	TOOLS	CLOCKS
COSMETICS	SOAP	PAINTS	BOXES
CELLOPHANE	FILM	OILS	CLOTHES

Name_____ Date_____

EVERYDAY LIFE SCIENCE

SEED PREDICTIONS

Many fruits contain seeds. In this activity, you will predict the number of seeds found in the fruits below. As a class, divide into teams. Each team should choose one of the fruits to bring in and cut open for the class. Record the number of seeds found in each piece of fruit.

FRUIT	PREDICTED NUMBER OF SEEDS	ACTUAL NUMBER OF SEEDS
Apple		
Pear		
Orange		
Lemon		
Grapefruit		
Cantaloupe		

In which fruit were you able to make your closest prediction? _____

In which fruit did you make your worst prediction? _____

Complete a class survey.

How many students made the closest prediction for the following:

_____ apple   _____ pear   _____ orange

_____ lemon   _____ grapefruit   _____ cantaloupe

everyday life science

SEEDS FROM TREE FRUITS

Many trees produce seeds. Obtain seeds from fruits that grow on the trees listed below. Wash, dry, and glue the seeds in the spaces below.

APPLE	GRAPEFRUIT	CHERRY	LEMON
ORANGE	ALMOND	DATE	NECTARINE
LIME	APRICOT	FIG	PEACH

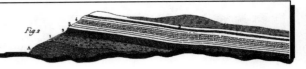

Roots and Shoots

Plants develop from seeds. In this activity, you will measure the length of the roots and shoots of a plant as it develops from its seeds.

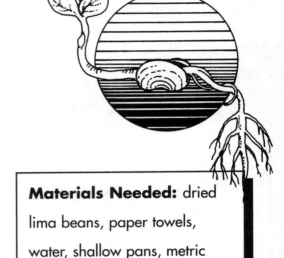

DIRECTIONS:

1. Soak some dried lima beans in a jar of water overnight.

2. Place some wet paper towels in the bottom of a shallow pan or dish.

3. Arrange five of the soaked lima beans across the middle of the pan.

4. Cover with several thicknesses of wet paper towels.

5. Wrap the pan with plastic food wrap to prevent rapid evaporation of the water.

Materials Needed: dried lima beans, paper towels, water, shallow pans, metric ruler, jar, plastic food wrap

6. Each day, carefully uncover the beans and observe any growth of roots or shoots. Once growth begins, measure the lengths of the roots and shoots in centimeters.

7. Keep a record of your measurements and observations in the chart below.

DAY	OBSERVATIONS	LENGTH OF ROOTS (CM)	LENGTH OF SHOOTS (CM)
1			
2			
3			
4			
5			
6			
7			

SPROUTING RADISHES

It's a lot of fun to watch plants grow from seeds. In this activity, you will discover the best conditions for the sprouting of radish seeds.

Materials Needed:

3 jar lids, salt water, fresh water, vinegar, paper towels, scissors, radish seeds

DIRECTIONS:

1. Cut six pieces of paper towel to fit inside the jar lids.

2. Place one piece of paper towel inside each lid.

3. Count out 25 radish seeds for each lid. Carefully place them inside each lid.

4. Cover the seeds in each lid with another piece of paper towel.

5. Pour enough fresh water inside the first lid to thoroughly wet the paper towel.

6. Pour enough salt water inside the second lid to thoroughly wet the paper towel.

7. Pour enough vinegar inside the third lid to thoroughly wet the paper towel.

8. Label each lid with the name of the liquid.

9. Set the lids aside for about three days.

10. After three days, carefully remove the top paper towel and observe the seeds.

11. Record your observations in the chart below.

LIQUID	NUMBER OF SEEDS SPROUTED	OBSERVATIONS
Fresh water		
Salt water		
Vinegar		

What conclusions can you make from these results?_____

_____

EVERYDAY LIFE SCIENCE

Name _____ Date _____

A TREE FROM A SEED

It's hard to believe that many of the big trees began as a seed. A tree seedling of an orange, lemon, lime, or grapefruit can be started in your very own classroom. Just follow the directions below.

DIRECTIONS:

1. Prepare a mixture of potting soil and sand.

2. Punch several small holes in the bottoms of four cups. Label the cups as follows: #1—orange, #2—lemon, #3—lime, and #4—grapefruit.

3. Use the soil mixture to fill the cups about two-thirds full. Do not pack the soil in the cups.

4. Remove the whole seeds from a freshly-cut orange.

5. Rinse the seeds in warm water.

6. Place several orange seeds in cup #1. Cover with one-half inch of the soil mixture. Water lightly and set in a warm place.

7. Repeat steps 4–6 with the lemon, lime, and grapefruit.

8. Check the cups each day. Do not allow the soil to dry out completely.

9. Complete the chart below to record your observations of how the seeds sprout.

> **Materials Needed:** fresh seeds from oranges, lemons, limes, and grapefruits; a knife; potting soil; sand; 4 paper or plastic foam cups; water

DATE	ORANGE	LEMON	LIME	GRAPEFRUIT

• • • • • • • • •
E V E R Y D A Y

life science

MANY TYPES OF OAK TREES

Trees belong to families. For example, oak trees belong to the beech family of trees. There are many types of oak trees. Unscramble the letters below to spell the names of some oak trees.

1. R A B N E R _____ _____ _____ _____ _____ _____

2. K A T B E S _____ _____ _____ _____ _____ _____

3. R E A B _____ _____ _____ _____

4. C L A B K _____ _____ _____ _____ _____

5. R U B _____ _____ _____

6. W O C _____ _____ _____

7. N I O R _____ _____ _____ _____

8. L E U A R L _____ _____ _____ _____ _____ _____

9. V I E L _____ _____ _____ _____

10. N I P _____ _____ _____

11. S T O P _____ _____ _____ _____

12. D R E _____ _____ _____

13. R A W T E _____ _____ _____ _____ _____

14. H E T I W _____ _____ _____ _____ _____

15. L O W L I W _____ _____ _____ _____ _____ _____

16. L O W L E Y _____ _____ _____ _____ _____ _____

17. N O S I O P _____ _____ _____ _____ _____ _____

EVERYDAY LIFE SCIENCE

EVERGREEN TREES OF THE USA

Eight evergreen trees of the United States will be identified in this activity. Find the names of these trees by decoding the words. The first number indicates the number of spaces across in the grid, and the second number indicates the number of spaces up. For example, the number 4–2 would be the letter W.

9	N	E	O	R	Y	M	O	U	E
8	C	A	H	E	D	E	S	N	A
7	A	S	N	L	I	F	N	K	M
6	R	E	R	L	N	U	B	A	L
5	P	C	O	R	M	P	L	D	Y
4	I	B	T	W	G	E	S	E	R
3	H	O	N	H	U	O	I	J	A
2	L	S	T	W	L	R	V	Y	H
1	F	R	N	E	T	Y	H	C	P
	1	2	3	4	5	6	7	8	9

1. ___ ___ ___ ___ ___ ___ ___ ___   ___ ___ ___ ___ ___
   1-7 5-5 9-9 6-2 5-7 1-8 9-3 1-9   9-2 6-3 9-6 1-2 6-1

2. ___ ___ ___ ___ ___ ___
   8-2 2-8 5-3 9-1 3-9 8-8

3. ___ ___ ___ ___ ___   ___ ___ ___
   2-7 4-2 2-9 4-1 5-1   2-4 8-6 9-5

4. ___ ___ ___ ___ ___ – ___ ___ ___ ___ ___ ___ ___ ___
   5-4 3-6 4-8 9-3 3-4   1-1 5-2 3-9 4-2 4-8 9-4 9-9 5-8

   ___ ___ ___ ___ ___ ___ ___ ___
   6-9 8-6 5-4 3-7 7-9 9-6 5-7 9-3

5. ___ ___ ___ ___ ___ ___ ___ ___ ___   ___ ___ ___ ___
   2-7 3-8 6-3 3-6 3-4 7-5 4-8 8-6 6-7   1-5 5-7 3-1 9-9

6. ___ ___ ___ ___ ___ ___ ___ ___   ___ ___ ___ ___
   9-6 6-3 7-6 1-2 3-5 7-5 9-6 8-2   9-1 7-3 5-6 8-4

7. ___ ___ ___   ___ ___ ___ ___ ___
   2-1 8-4 5-8   2-5 8-4 5-8 9-3 6-2

8. ___ ___ ___ ___ ___ ___   ___ ___ ___ ___ ___ ___ ___
   2-5 6-3 6-9 9-7 2-3 3-3   8-3 5-3 3-1 5-7 1-5 9-9 2-1

FS-10616 Everyday Life Science

everyday life science

THE ROSE FAMILY

Many favorite fruits grow on trees that belong to the rose family. The springtime blooms on these trees create beautiful scenes in parks, gardens, and yards.

Use reference books or go outside to observe the appearance, color, and aroma of the fruits below of the rose family.

CHERRY	
PLUM	
PEACH	
PEAR	
APPLE	

What are the main similarities in all these fruits? _____

_____

_____

What are the main differences in these fruits? _____

_____

_____

List some food products that contain these fruits. _____

_____

EVERYDAY *life science*

Can You Identify the Genus?

The first word of a two-part scientific name used to classify a group of closely-related species is called the genus. Carolus Linnaeus, a Swedish botanist, developed a classification system that is used by scientists to classify things. Match the genus of each tree below with its common name.

1. _____ Pinus	**A.** willow
2. _____ Sequoia	**B.** hickory
3. _____ Juniperus	**C.** elm
4. _____ Ginkgo	**D.** magnolia
5. _____ Salix	**E.** poplar
6. _____ Populus	**F.** cherry
7. _____ Juglans	**G.** redwood
8. _____ Carya	**H.** maple
9. _____ Quercus	**I.** pear
10. _____ Ulmus	**J.** holly
11. _____ Magnolia	**K.** pine
12. _____ Pyrus	**L.** ginkgo
13. _____ Prunus	**M.** dogwood
14. _____ Ilex	**N.** walnut
15. _____ Acer	**O.** oak
16. _____ Cornus	**P.** juniper
17. _____ Fraxinus	**Q.** ash

Which of the genuses were the easiest to identify? _____

_____

_____

Which of the genuses were the most difficult to identify? _____

_____

_____

TREE FOODS

Trees provide people with many wonderful foods. Choose one of the foods below. Complete the chart to show the different uses of and products made using the food. Then design a poster that displays the different uses and products of the food.

Food containers and food wrappers which list the selected food as an ingredient can be used on the poster. Bring a food product containing the food to school to help create a class display.

APPLE	ORANGE	LEMON	LIME	GRAPEFRUIT
PEAR	PEACH	PLUM	COCONUT	PECAN
ALLSPICE	ALMOND	APRICOT	AVOCADO	CACAO
CASHEW	CHERRY	CHESTNUT	CINNAMON	CLOVE
DATE	FIG	MANGO	NECTARINE	NUTMEG
OLIVE	PECAN	PISTACHIO	PRUNE	WALNUT

FOOD _____

USES OF	PRODUCTS

EVERYDAY LIFE SCIENCE

EVERYDAY LIFE SCIENCE

A COLLECTION OF LEAVES

Leaves come in all shapes and sizes depending on the type of tree on which they grow. Prepare a poster of leaves using the labels below. Collect several samples of leaves which match each label. Place the sample leaves between layers of paper towels, and press them flat with heavy books or bricks for several days. Attach the labels and examples of the leaves to a sheet of posterboard. Display your leaf collection in class. Compare your collection with other students' collections.

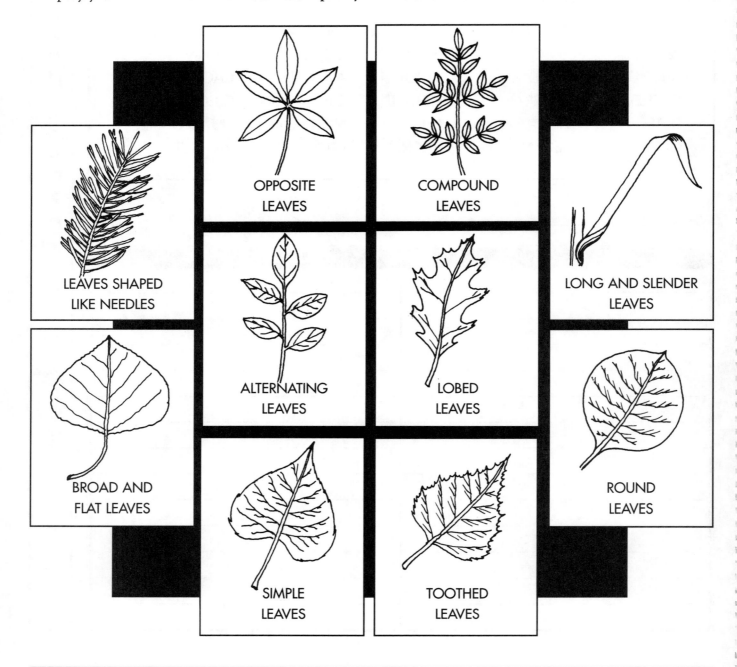

Name _____ Date _____

PARTS OF A FLOWER

Most flowers have four main parts. Each of these parts consists of elements. These parts and elements are listed below. In this activity, you will display some of the parts of a real flower.

Materials Needed: one of the following kinds of flowers: daffodils, roses, lilies, tulips, irises, and phlox make good samples; sharp cutting tool (adult supervision and help is required); paper towels; heavy books or bricks; construction paper or posterboard; glue; black marker

1. After you have collected a good flower sample, have an adult help you make a cross-section of the flower. This can be done by carefully slicing the flower using a sharp cutting tool.

2. Place the cross-section between several sheets of paper towels and press for several days with heavy books or bricks.

3. Carefully remove the paper towel. Attach the pressed flower to a sheet of construction paper or posterboard.

4. Cut out the labels to the right. Attach them to the construction paper or posterboard. Use a black marker to draw lines from the labels to the parts of the flower.

PETALS	ANTHER
POLLEN GRAINS	STIGMA
STYLE	STAMENS
FILAMENT	COROLLA
SEPALS	CALYX
OVARY	OVULE
PISTILS	

EVERYDAY LIFE SCIENCE

LEAF NIBBLES

Insects and other organisms often cause much damage to leaves. In this activity, you will examine leaves for evidence of damage from insects and other organisms. Collect one leaf for each of the four types of damage listed below. Attach each leaf to the appropriate labeled section.

Leaf with chewed edges

Leaf with holes chewed in it

Leaf on which only one side has been eaten

Leaf on which a network of veins is all that remains

Name_____   Date _____

PLANT FOODS

People eat all kinds of plants for food. Plant foods come in many shapes, colors, odors, tastes, and sizes. Use the word clues below and the letter grid to find a hidden message about some plants people eat.

1. A good example of a food from the leaves of a plant is
   a. a tomato—Darken all the A's in the grid.
   b. a cabbage—Darken all the G's in the grid.
   c. a potato—Darken all the H's in the grid.

2. A good example of a food from the stems of a plant is
   a. celery—Darken all the J's in the grid.
   b. a strawberry—Darken all the L's in the grid.
   c. a radish—Darken all the M's in the grid.

3. A good example of a food from the roots of a plant is
   a. beans—Darken all the N's in the grid.
   b. asparagus—Darken all the P's in the grid.
   c. peanuts—Darken all the W's in the grid.

4. A good example of a food from the flowers of a plant is
   a. cauliflower—Darken all the K's in the grid.
   b. sugarcane—Darken all the S's in the grid.
   c. carrot—Darken all the T's in the grid.

5. A good example of a food that is called a fruit is
   a. sweet potato—Darken all the V's in the grid.
   b. spinach—Darken all the X's in the grid.
   c. apricot—Darken all the Q's in the grid.

G	P	W	L	J	A	K	G	N	Q	T	Q	J	S	W	G	A	R	K	E
W	G	A	J	K	W	N	Q	G	E	J	X	C	G	E	L	L	E	N	T
J	S	Q	G	O	K	W	U	Q	J	G	Q	R	K	W	J	C	G	Q	E
G	W	O	J	K	W	F	G	Q	J	M	I	W	N	E	G	R	A	L	S
A	G	J	N	D	K	V	G	I	W	T	Q	A	K	J	M	I	G	N	S
J	F	K	G	W	O	J	Q	R	K	G	Q	W	J	Q	A	K	J	W	G
H	G	Q	E	J	G	W	A	K	G	Q	L	J	W	T	G	H	K	J	Y
G	B	J	Q	W	G	O	K	W	Q	J	D	K	J	Q	W	Y	J	K	Q

Use the remaining letters to write out the hidden message below:

_____ _____ _____ _____

_____ _____ _____ _____ _____

_____ _____.

EVERYDAY life science

Pigment Art

For thousands of years, people have used the natural pigments from berries, flowers, bark, leaves, and roots of plants to create artwork and dyes for fabrics. Collect a variety of leaves, flowers, peels, roots, stems, and berries from plants to create your own drawing below. To do this, rub the plant parts on the paper. At the bottom of the drawing, list the sources of the colors you used in your drawing. Display the artwork on a bulletin board or in a class book. Make a class list of the plants used to produce different shades of red, orange, yellow, green, blue, or violet.

Sources of plant pigments: _____

_____

_____

HIDDEN COLORS IN FLOWER PIGMENTS

Chromatography is the process of separating the colors in pigments. Some pigments are leaves, flower petals, inks, or dyes. To understand how this process works, try the experiment below.

1. Select one type of flower to use in this experiment. Flowers with red, blue, or purple petals will give the best results. Remove some of the petals.

2. Flatten two coffee filters on a sheet of newspaper.

3. Rub some of the flower petals in the center of each coffee filter in order to produce a smudge mark of pigment.

> **Materials Needed:** paper
> coffee filters, newspapers,
> eyedroppers, flower petals,
> vinegar, household ammonia

4. With an eyedropper, add several drops of vinegar to the center of one filter paper. Add several drops of household ammonia to the center of the other filter paper.

5. Observe for 5–10 minutes. Record your observations in the chart below.

Type of flower used in the experiment: _____

OBSERVATIONS WITH VINEGAR	OBSERVATIONS WITH AMMONIA

What were some of the differences in the color separations in the two trials?

_____

_____

EVERYDAY LIFE SCIENCE

EVERYDAY LIFE SCIENCE

A FUNGUS AMONG US

A fungus is a simple nongreen plant. It lacks chlorophyll, a green substance plants use to make food. Draw one type of fungus following the directions below.

1. Plot the coordinate points in the grid. The first number represents the horizontal line, and the second number represents the vertical line. The first point has been done for you.

2. Connect the points with straight lines in the order of plotting.

3. When you have completed the pattern, identify the type of fungus you have drawn.

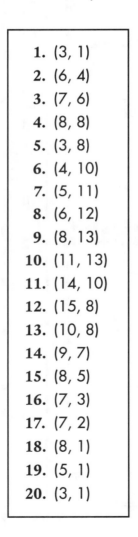

1. (3, 1)
2. (6, 4)
3. (7, 6)
4. (8, 8)
5. (3, 8)
6. (4, 10)
7. (5, 11)
8. (6, 12)
9. (8, 13)
10. (11, 13)
11. (14, 10)
12. (15, 8)
13. (10, 8)
14. (9, 7)
15. (8, 5)
16. (7, 3)
17. (7, 2)
18. (8, 1)
19. (5, 1)
20. (3, 1)

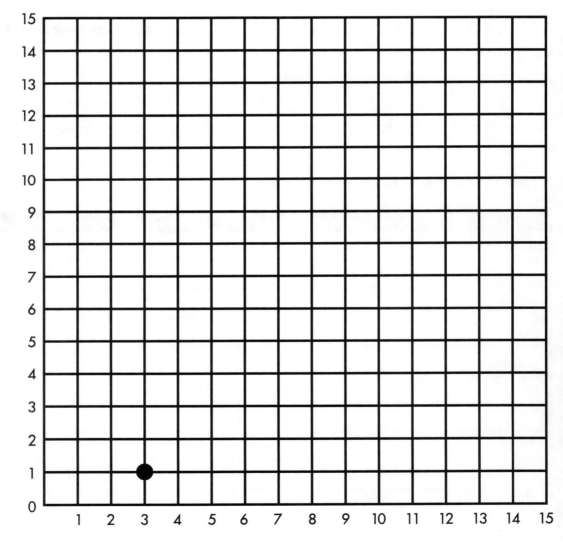

Identify the fungus you have sketched. _____

FUNGI GROWTH

Fungi are simple nongreen plants. They contain no chlorophyll, a substance that enables plants to make their own food. In the activity below, you will observe the conditions which permit the most rapid growth of fungi.

Materials Needed:
6 resealable sandwich bags, 6 slices of white bread, water, vinegar

DIRECTIONS:

1. Place one slice of dry white bread in a sandwich bag and close. Label the bag "**A.**" Place the bag in a bright sunny place.

2. Place one slice of dry white bread in a sandwich bag and close. Label the bag "**B.**" Place the bag in a dark place.

3. Sprinkle one slice of bread with water. Place it in a bag and close. Label the bag "**C.**" Place the bag in a bright sunny place.

4. Sprinkle one slice of bread with water. Place it in a bag and close. Label the bag "**D.**" Place the bag in a dark place.

5. Sprinkle one slice of bread with vinegar. Place it in a bag and close. Label the bag "**E.**" Place the bag in a bright sunny place.

6. Sprinkle one slice of bread with vinegar. Place it in a bag and close. Label the bag "**F.**" Place the bag in a dark place.

7. Examine the bags each day for one week. Record your observations in the chart below.

Day	Bag A	Bag B	Bag C	Bag D	Bag E	Bag F
1						
2						
3						
4						
5						
6						
7						

CONCLUSIONS: What conditions produced the most mold on the bread?

_____

What effect did the vinegar have on the production of mold?

_____

Name_____ Date _____

EVERYDAY LIFE SCIENCE

YEAST ACTION

Fungi organisms called yeasts are used in bread making. In this activity, you will observe the conditions that cause the yeasts to react best.

1. Add one package of dry yeast and 200 mL of warm water to a freezer bag. Swirl gently to dissolve the yeast. Add 2 tablespoons of sugar to the bag, seal the bag, and swirl gently to mix the sugar. Set aside in a warm place.

2. To the second freezer bag, add one package of dry yeast and 200 mL of warm water. Swirl gently to dissolve the yeast. Add 10 tablespoons of flour to the bag, seal the bag, and swirl gently to mix the flour. Set aside in a warm place.

3. To the third freezer bag, add one package of dry yeast and 200 mL of warm water. Swirl gently to dissolve the yeast. Add 2 tablespoons of sugar and 10 tablespoons of flour to the bag, seal the bag, and swirl gently to mix the sugar and flour. Set aside in a warm place.

4. Observe the reactions in the three bags over a one-hour period. Record your observations in the chart below.

OBSERVATIONS:

TIME	BAG 1— YEAST AND SUGAR	BAG 2— YEAST AND FLOUR	BAG 3— YEAST, SUGAR, AND FLOUR
15 min.			
30 min.			
45 min.			
60 min.			

Which trial produced the best results that account for the rising of bread?

_____

everyday life science

HOW MANY ORGANISMS LIVE HERE?

An organism is an individual form of life. Organisms include plants, animals, and microscopic living things. To learn more about organisms, try the activity below. In this activity, you will observe the different types and numbers of organisms in a selected, outdoor study site.

1. Locate an outdoor study site that is shaded and that contains loose, rich soil.

2. Using the meterstick, mark off a one-square meter area. Place the sliced potatoes on top of the soil throughout the area.

Materials Needed: meterstick, potato slices, wet newspapers or cardboard, a magnifying lens, rocks or boards

3. Cover the area completely with wet newspapers or cardboard. Place rocks or boards on the corners to keep the paper in place.

4. Do not disturb the study area for several days.

5. After several days, return to the study site with the chart below, a pencil, and a magnifying lens.

6. Carefully remove the rocks or boards and the paper.

7. Identify the different types and numbers of organisms found.

Describe the area you chose for your study site. _____

TYPE OF ORGANISMS FOUND	NUMBER OF THESE ORGANISMS

Compare your findings with those of your class members.

FS-10616 Everyday Life Science

EVERYDAY **life science**

Harmful Effects

Plants, animals, and people have suffered from the effects of small organisms. The health and economy of large regions of Earth have been affected by the events listed below. Select an event below or one of your own to research.

Record some of the chosen event's main characteristics in the chart below. This can be attached to a bulletin board. Prepare a report in greater detail to present to the class.

DUTCH ELM DISEASE   FIRE ANT INFESTATION   CHOLERA

CHESTNUT BLIGHT   MALARIA   E. COLI FOOD POISONINGS

MUSEUM BEETLE INFESTATION   YELLOW FEVER   PNEUMONIA EPIDEMICS

Name of Chosen Event: _____

Special Characteristics: _____

_____

_____

_____

_____

Effects on Other Living Things: _____

_____

_____

_____

_____

Other: _____

_____

_____

_____

_____

Name_____ Date_____

OUR ENVIRONMENT

It is very important to take care of and protect our environment. To learn some ways to do this, use the words from the word list and the clues below to complete the crossword puzzle about our environment.

AIR	NOISE
ANIMALS	OFF
AWAY	PEOPLE
CARS	PLANTS
EARTH	RAGS
ENVIRONMENT	TREES
EVERYONE	TURN
FISH	UGLY
GIVE	WASTE
HURTS	WATER
INSECTS	WATER
LITTER	

ACROSS

1. "Every Litter Bit _____!"
3. Trash where it doesn't belong
8. What every living thing needs for life
9. These are good to clean with.
13. _____ needs to help the environment!
14. Something needed by each living thing in the environment
16. Many are killed by dirty water that comes from factories.
18. Provide oxygen for the air
20.-21. When you run the shower too long, you _____ _____.
22. Opposite of beautiful
23. These emit poisonous gases into the air.

DOWN

2. and 11. _____ _____ the lights when you leave a room.
4. Raw material for paper
5. The planet that makes up our total environment
6. Many _____ have had their habitats destroyed by people.
7. _____ pollution is harmful to a person's ears.
10. and 19. _____ _____ good used items rather than destroying them.
12. Too many _____ living in one place have harmed the environment.
15. Needed by all living things
17. Eaten by most birds

Name _____   Date _____

DECOMPOSITION OF FOODS

Food decomposes at different rates depending on the type of container in which it is kept. In this activity, you will observe the rate of decomposition of certain foods in different types of containers.

| **Materials Needed:** pieces of banana, carrot, potato, bread, lettuce, and crackers; one 30-cm square piece of each of the following: plastic food wrap, newspaper, cloth; 3 rubber bands; optional: flowerpots filled with soil |

DIRECTIONS:

1. Place equal amounts of all of the food pieces in the center of each of the three types of material.

2. Make each square of material into a pouch and fasten it tightly with a rubber band.

3. Bury each pouch the same depth in flowerpots filled with soil from outside your classroom or home.

4. After one week, dig up the pouches. Open them carefully. Examine the contents and record your observations in the chart below.

5. Close the pouches and bury them again in the soil.

6. At the end of the second week, dig up the pouches, open them, and examine the contents. Record your observations in the chart below.

POUCH	OBSERVATIONS AFTER WEEK 1	OBSERVATIONS AFTER WEEK 2
plastic		
paper		
cloth		

Describe the differences in the contents of the three pouches after two weeks. _____

_____

_____

Which wrapping provided the best protection from decomposition? _____

Which wrapping provided the poorest protection from decomposition? _____

     FS-10616 Everyday Life Science

AIR POLLUTANTS

People all over the world pollute Earth's air with millions of kilograms of "aerial garbage" each year. The health of plants and animals is greatly affected by these pollutants.

Unscramble the letters below to identify some of these pollutants. Then match each pollutant to its description.

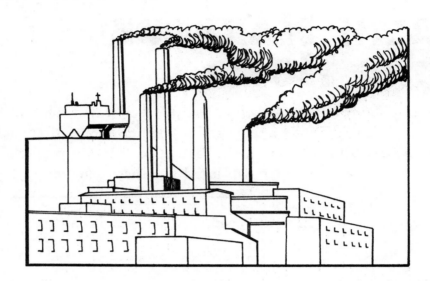

____ **1.** B R A N O C   D I X M O O N E

_____   _____

____ **2.** S T I L A P A R C U T E

_____

____ **3.** T R E G N I O N   X I D I O D E

_____   _____

____ **4.** S N A B R O C H D O R Y

_____

____ **5.** H O M I C P E T O C H L A   G O M S

_____   _____

____ **6.** F R U L U S   I D E O X I D

_____   _____

A. This poisonous gas comes from car exhaust. It causes people to be dizzy and have headaches.

B. These poisonous gases come from factories and power plants that burn coal or oil containing sulfur. They harm our respiratory systems.

C. Nitrogen dioxide combines with hydrocarbons and sunlight to create smog.

D. Unburned chemicals in combustion, such as car exhaust, react in air to produce smog.

E. Smoke, ash, dust, fumes, and other solid particles in the air

F. Mixture of gases and particles oxidized by the sun from products of gasoline and other burning fuels

EVERYDAY LIFE SCIENCE

HEALTH HAZARDS IN THE HOME

Homes can be dangerous places when you stop to think of all the things in them that can potentially harm you if used improperly. Take a survey in your home for the materials listed below that contain chemicals that could be poisonous. After your survey is complete, discuss the proper handling and storage of these materials with your classmates.

_____ aspirin	_____ nail polish	_____ perfume	_____ lotions
_____ ointment	_____ rubbing alcohol	_____ cough medicine	_____ shampoo
_____ laundry detergent	_____ dish detergent	_____ bleach	_____ shoe polish
_____ deodorizers	_____ house plants	_____ paints	_____ paint thinner
_____ kerosene	_____ gasoline	_____ glue	_____ putty
_____ crayons	_____ turpentine	_____ mineral spirits	_____ lead solder
_____ fertilizers	_____ household ammonia	_____ lighter fluid	_____ berries
_____ flowers	_____ shrubs	_____ pesticides	_____ vines
_____ wildflowers	_____ mushrooms	_____ antifreeze	_____ acids

List other materials you find.

_____

_____

_____

_____

_____

_____

_____

_____

_____

everyday life science

CONCEPT MAPPING AND THE HUMAN BODY

A concept map is a graphic representation of the key terms or ideas that relate to a central theme. Branches and sub-branches are used to show how the terms are organized into different groups.

On a separate sheet of paper, construct your own concept map following the diagram. With the term HUMAN BODY as the central theme, identify the five terms below that represent five subtopics. Cut them out and glue those five terms to the five branches. Then cut out and glue the other terms to their subtopics. Lines may be drawn to show the connections.

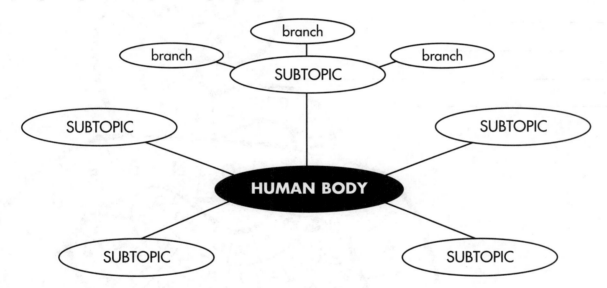

collar bones	intestines	testes	skeletal system	organs
circulation	smallpox	diseases	endocrine system	skull
pelvis	liver	adrenal gland	shoulder blades	legs
spinal column	measles	movement	body functions	rabies
stomach	thyroid	respiration	hepatitis	hands
parathyroid	influenza	excretion	thymus	neck
heart	digestion	tetanus	colds	feet
tuberculosis	lungs	arms	reproduction	sternum
nerve reactions	pituitary	esophagus	gall bladder	ovaries

EVERYDAY life science

The Human Digestive System

Important parts of the human digestive system are hidden in the maze below. From the starting letter, skip every other letter and write the words in the spaces.

1. ___ ___ ___ ___ ___ ___

2. ___ ___ ___ ___ ___ ___ ___ ___ ___

3. ___ ___ ___ ___ ___ ___ ___

4. ___ ___ ___ ___ ___ ___

   ___ ___ ___ ___ ___ ___ ___

5. ___ ___ ___ ___ ___

   ___ ___ ___ ___ ___ ___ ___

6. ___ ___ ___ ___ ___ ___

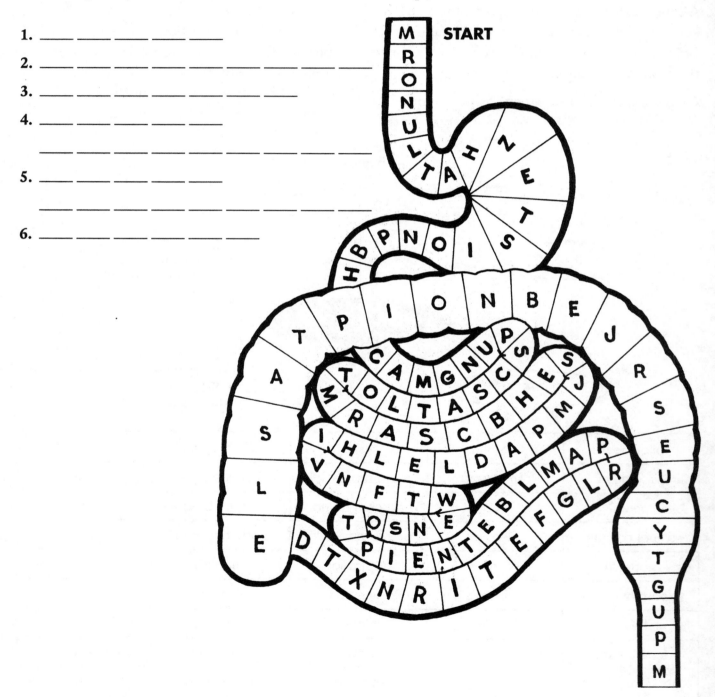

MUSCLE MANIA

Muscles are tough elastic tissues that enable body parts to move. Changes in temperature can affect the muscles. To find out how, try the experiment below at home, and share your results with your class members the next day.

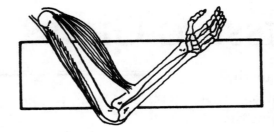

Materials Needed: a bowl large enough to place your hand in, ice cubes, water, warm water, paper towels, watch or clock with a secondhand, pencil

1. Write or print your name 10 times on the back of this page.

2. Place some ice cubes and water in the bowl. Hold your writing hand in the water for 30 seconds. Dry your hand quickly with a paper towel. Write or print your name 10 times in the spaces below.

_____   _____

_____   _____

_____   _____

_____   _____

_____   _____

3. Replace the ice water with warm water. Place your writing hand in the warm water for 30 seconds. Dry your hand quickly with a paper towel. Write or print your name 10 times in the spaces below.

_____   _____

_____   _____

_____   _____

_____   _____

_____   _____

4. Compare the appearance of your handwriting. How did the changes in temperature affect your handwriting samples?_____

_____

5. How does temperature affect your muscles? _____

_____

© Frank Schaffer Publications, Inc.

FS-10616 *Everyday Life Science*

EVERYDAY LIFE SCIENCE

EVERYDAY LIFE SCIENCE

MISSING GENES

Genes are the parts of cells that determine the characteristics that living things inherit from their parents. To learn about some important terms relating to genes, circle the word GENE anytime it appears in the puzzle. The remaining letters will spell out words relating to the study of heredity. Print those words in the spaces at the bottom of the page.

```
T  G  E  N  E  R  G  A  I  T  S  G  E  N  E  D
O  E  M  I  N  G  E  N  E  A  N  T  G  E  N  E
H  N  E  R  E  E  N  D  I  T  Y  G  E  N  E  H
Y  E  B  G  E  N  E  R  I  D  G  E  N  E  I  N
H  E  G  E  N  E  R  I  T  G  E  N  E  A  N  C
E  G  E  N  E  O  F  F  G  E  N  E  S  P  R  I
N  G  G  E  N  E  G  G  E  N  E  E  N  E  T  I
G  E  N  E  C  S  G  E  N  E  C  H  R  G  O  G
G  E  N  E  M  O  E  G  E  N  E  S  G  E  G  E
E  O  M  E  G  E  N  E  Z  Y  G  G  E  N  E  N
N  O  G  E  N  E  E  N  T  E  M  E  N  E  N  E
E  N  E  D  E  L  G  E  N  E  R  E  E  C  E  E
G  E  N  E  S  S  I  G  E  N  E  G  E  N  E  V
E  G  E  N  E  P  A  R  E  N  T  S  G  E  N  E
```

1. __ __ __ __ __ __

2. __ __ __ __ __ __ __ __

3. __ __ __ __ __ __ __ __

4. __ __ __ __ __ __

5. __ __ __ __ __ __ __ __ __ __ __

6. __ __ __ __ __ __ __ __

7. __ __ __ __ __ __ __

8. __ __ __ __ __ __ __

9. __ __ __ __ __

10. __ __ __ __ __ __

11. __ __ __ __ __ __

12. __ __ __ __ __ __

A MAGIC SQUARE OF CELLS

• • • • • • • • • •
E V E R Y D A Y

life science

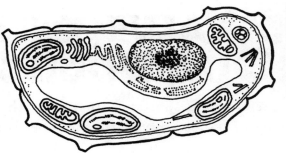

A cell is the basic unit of life. All living things are made up of cells. To learn about plant cells, read the clues concerning the parts of a plant cell in the boxes below. Then select the correct term from the word list that applies to each clue, and write the number for that term in the box.

By recording all the correct numbers, you have produced a magic square. When you add the numbers across, down, or diagonally, you should get the same answer.

1. cell membrane

2. nucleus

3. chromosomes

The membrane-enclosed command center of the plant cell	The rigid layer that supports and protects the cell	The gel-like substance in which most of the cell's life processes take place
_____	_____	_____

4. mitochondria

5. chloroplasts

6. cytoplasm

Places where cells store water, food, and other materials	Places where cells convert light energy into food	The outer covering of the cell
_____	_____	_____

7. cell wall

8. nucleolus

9. vacuoles

These supply energy needed by the cell to do work.	These contain complex chemical information that directs the cell's hereditary-related activities.	A region in the nucleus that produces tiny cell particles needed in protein synthesis
_____	_____	_____

What is your numerical answer? _____

 FS-10616 Everyday Life Science

EVERYDAY LIFE SCIENCE

THE VALUE OF MINERALS IN OUR BODIES

Minerals are inorganic substances that aid in chemical reactions and serve as building materials in our bodies. They also control the movement of fluids in and out of the cells.

Locate a chart on nutrients in your text or other reference and complete the chart.

MINERAL	CHEMICAL SYMBOL	FOOD SOURCES	HOW USED BY THE BODY
iron			
calcium			
iodine			
phosphorus			
potassium			
sodium			
magnesium			
manganese			
copper			
sulfur			

Attach two food labels which list some of these minerals to the back of this page.

everyday life science

THE FOOD GUIDE PYRAMID

The Food Guide Pyramid is a guide you can use to help you eat healthy every day. For this activity, work in a team of six students. Reproduce the Food Guide Pyramid on a large sheet of posterboard or butcher paper. Using magazines and newspapers, each team member should find and cut out pictures of foods for one of the sections on the pyramid. Glue these pictures to the chart.

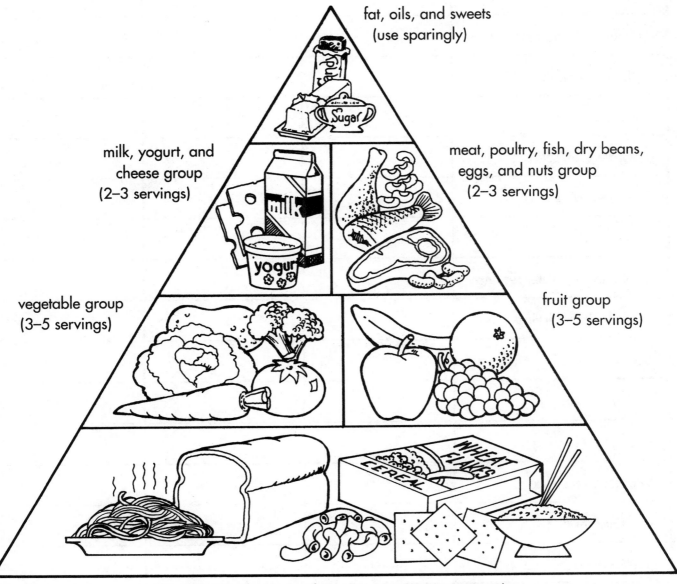

fat, oils, and sweets
(use sparingly)

milk, yogurt, and
cheese group
(2–3 servings)

meat, poultry, fish, dry beans,
eggs, and nuts group
(2–3 servings)

vegetable group
(3–5 servings)

fruit group
(3–5 servings)

bread, cereal, rice, and pasta group (6–11 servings)

EVERYDAY life science

Nutrition Facts About Crackers

Some foods are better for you than others. In this activity, you will examine the labels of three different brands of saltine, cheese, or graham crackers for nutritional data. Complete the chart below to determine which of the crackers is the most healthy for you.

	BRAND A	BRAND B	BRAND C
Brand name of product			
Number of crackers/serving			
Calories/serving			
Total fat/serving			
Saturated fat/serving			
Cholesterol/serving			
Sodium/serving			
Total carbohydrate/serving			
Dietary fiber/serving			
Sugars/serving			
Protein/serving			
Vitamin A/serving			
Vitamin C/serving			
Calcium/serving			
Iron/serving			

NUTRITION FACTS ABOUT CEREALS

Some cereals have more vitamins in them than others. In this activity, you will examine the labels of three different brands of cereal to compare the nutritional data. Complete the chart below for cereal without milk to determine which of the cereals is the most healthy for you.

	BRAND A	BRAND B	BRAND C
Brand name of cereal			
Serving size			
Calories			
Total fat			
Saturated fat			
Cholesterol			
Sodium			
Potassium			
Total carbohydrate			
Dietary fiber			
Sugars			
Protein			
Vitamin A			
Vitamin C			
Calcium			
Iron			
Vitamin D			
Thiamin			
Riboflavin			
Niacin			
Phosphorus			
Magnesium			
Zinc			

EVERYDAY LIFE SCIENCE

EVERYDAY LIFE SCIENCE

BACTERIA OR VIRUSES

Disease is any uneasiness, distress, or discomfort to any part of the body. Many infectious diseases are caused by an invasion of the body by viruses or bacteria. The word *virus* has a Latin origin meaning *poison*.

The word *bacterium* has a Greek origin meaning *rod*.

Identify the causes of the following human infectious diseases. Place V for virus or B for bacteria in the space before each disease.

1. _____ ear infection
2. _____ tooth decay
3. _____ croup
4. _____ whooping cough
5. _____ chickenpox
6. _____ meningitis
7. _____ food poisoning
8. _____ mumps
9. _____ hepatitis
10. _____ influenza
11. _____ shingles
12. _____ rabies
13. _____ pneumonia
14. _____ encephalitis

15. _____ smallpox
16. _____ strep throat
17. _____ diphtheria
18. _____ scarlet fever
19. _____ impetigo
20. _____ tuberculosis
21. _____ tetanus
22. _____ measles
23. _____ HIV/AIDS
24. _____ common cold
25. _____ poliomyelitis
26. _____ cold sores
27. _____ yellow fever
28. _____ cholera

How are these diseases prevented and/or treated?

Those caused by bacteria: _____

_____

_____

Those caused by viruses: _____

_____

_____

HUMAN DISEASES

Diseases are common all over the world. Find the names of human diseases in the puzzle below. The words may be written up, down, forward, or backward. Circle the words.

CHICKENPOX	MEASLES	SMALLPOX
COMMON COLD	MUMPS	STREP THROAT
DIPHTHERIA	PNEUMONIA	TETANUS
HEPATITIS	POLIO	WHOOPING COUGH
INFLUENZA	RABIES	YELLOW FEVER

```
R A B E S U S U N A T E T C
A C L W Z N A C U B A R L T
T I D I P H T H E R I A N U
H G U O C G N I P O O H W B
I L O M T S A C V N L D C S
R A P R I P C K A F J N G I
A M H O S M S E L S A E M T
B I N F L U E N Z A K T L I
I P I U R M T P O L I O B T
E Y A C A I N O M U E N P A
S M A L L P O X D G K Z F P
E Y E L L O W F E V E R L E
D L O C N O M M O C I N O H
H J M T A O R H T P E R T S
```

EVERYDAY LIFE SCIENCE

PUZZLING DISEASES

Connect the letters in the correct order in each puzzle below to spell out the name of a disease.

1. Z A L F
N U N
E I

2. H C O X
I C P
K N
E

3. E M Y L
S S E
D A E
I S

4. S S A I
I P R
S O

5. A E S Y
M M H
E P

6. E R I A
H I D
T P
H

7. R E Y E L
V E W L
F O

8. B U T S I
E R L S
C U O

9. O M E P
N U N
I A

10. R E P A
E V R R
F T O

11. E V E R T
F I H Y
D O P

12. E U P C I
G L O N
A U B
B

▲▲▲▲▲▲▲▲▲▲▲▲▲▲▲▲▲▲▲▲▲▲▲▲▲▲▲▲▲▲▲▲▲▲

everyday life science

APPLE REACTIONS

You can compare how an apple reacts to disinfectants to how human skin reacts to disinfectants by completing the activity below. In this activity, you will observe the changes in cut and bitten apples after treatments with disinfectants.

> **Materials Needed:** cotton swabs, a knife, labels, two apples, soap and water, iodine solution, rubbing alcohol

1. Cut one apple into quarters. Label each slice as follows:
   A. Set one piece aside to serve as a control.
   B. Wash one slice with soap and water.
   C. Rub iodine solution on one slice.
   D. Rub rubbing alcohol on one slice.

2. Take four large bites out of the second apple but do not chew them. Label each piece as follows:
   E. Set one bite of apple aside to serve as a control.
   F. Wash one bite of apple with soap and water.
   G. Rub iodine solution on one slice.
   H. Rub rubbing alcohol on one slice.

Observe the changes in the apple slices and bites over a period of 10 days. Record your observations in the chart.

DAY	APPLE A	APPLE B	APPLE C	APPLE D
1				
2				
3				
4				
5				
6				
7				
8				
9				
10				

DAY	APPLE E	APPLE F	APPLE G	APPLE H
1				
2				
3				
4				
5				
6				
7				
8				
9				
10				

What were the different effects of the disinfectants on the cut apple and the bitten apple?

_____

How do disinfectants help the human skin from cuts, scratches, and bites?

_____

Name_____ Date_____

How Exercise Affects Your Pulse Rate

Exercise is important to keep your body healthy. But what happens to your pulse rate after you exercise? To find out, try the activity below. You will measure your pulse rate before and after a short exercise. You will need a clock with a secondhand.

1. Locate your pulse at your wrist with your forefinger and your middle finger. Press gently on the area where you can best feel your pulse.

2. Using a clock with a secondhand, count your pulse rate for 30 seconds. Record your result in the chart to the right.

TRIAL	PULSE RATE FOR 30 SECONDS	PULSE RATE FOR ONE MINUTE
1		
2		
3		
4		
5		

3. Repeat the pulse count four more times. Record your results in the chart.

4. Multiply each result by two to calculate your pulse rate for one minute.

5. Run in place for one minute. Then quickly find your pulse and take the count for 30 seconds. Record your result in the chart to the right.

6. Repeat the one-minute runs four more times. After each run, take your pulse count for 30 seconds, and record the result. Be sure to double each count and record your pulse rate for one minute.

TRIAL	PULSE RATE FOR 30 SECONDS	PULSE RATE FOR ONE MINUTE
1		
2		
3		
4		
5		

7. Calculate your average pulse rate before exercise._____

8. Calculate your average pulse rate after exercise. _____

9. How did exercise affect your pulse rate? _____

_____

STIMULI AND RESPONSES

How do you react when you touch something hot or step on a piece of broken glass? Your reflexes cause you to move quickly to protect yourself. The act of touching something hot is called the stimulus, and the act of moving quickly away from the hot object is called the response.

List below common stimuli and responses that affect your senses of touch, sight, smell, hearing, or taste. Share your list with class members.

STIMULUS	RESPONSE

© Frank Schaffer Publications, Inc.

FS-10616 Everyday Life Science

Name_____   Date_____

INHERITED BEHAVIOR

Inborn, or innate, behavior of an organism is inherited from its parents. The migration of animals or the building of nests are good examples of innate behavior. Inborn behavior includes reflexes and instincts.

List some inherited behaviors of the animals listed below.

ANIMALS	INHERITED BEHAVIORS
worms	
spiders	
insects	
fish	
snakes	
frogs	
turtles	
birds	
cats	
humans	

LEARNED BEHAVIOR

• • • • • • • •
E V E R Y D A Y

life science

Read about Ivan Pavlov's famous stimulus-response experiments he did with dogs in Russia in the early 1900s. The reaction of the dogs was a learned behavior. Swimming, riding a bicycle, or driving a car are learned behaviors.

In the chart, list some learned behaviors the animals do.

ANIMALS	LEARNED BEHAVIORS
dogs	
cats	
horses	
birds	
circus animals	
dolphins	
elephants	
chimpanzees	
cows	
humans	

EVERYDAY LIFE SCIENCE

INTERVIEW A LIFE SCIENTIST

Life scientists provide valuable contributions to society. Identify and contact one of the life scientists listed below that lives in your community. Ask the scientist for permission for an interview, and set up an appointment. Also, ask for permission if you plan to record the interview.

Prepare many of your questions before the interview. Questions may pertain to the person's interest in the area of science, when

that interest first developed, what special training is needed, what services are offered, what special equipment and supplies are needed, and what some new developments are in that area of science.

After completing the interview, prepare a written report you can share with your class.

Some of the areas of life science you might be interested in are as follows:

BIOLOGIST	NURSE	TREE SURGEON
BOTANIST	INTERNAL MEDICINE DOCTOR	ORTHOPEDIC SURGEON
BIOCHEMIST	FAMILY PRACTICE DOCTOR	PEDIATRICIAN
ZOOLOGIST	ONCOLOGIST	DENTIST
FORESTER	SURGEON	DENTAL HYGIENIST
HORTICULTURIST	DERMATOLOGIST	MEDICAL TECHNOLOGIST
MICROBIOLOGIST	PLASTIC SURGEON	ENTOMOLOGIST
VETERINARIAN	WILDLIFE MANAGER	ZOO CURATOR
ORNITHOLOGIST	ICHTHYOLOGIST	HEALTH SCIENTIST
BEE KEEPER	WILDFLOWER EXPERT	MAMMALOGIST
HERPETOLOGIST	X-RAY TECHNICIAN	ECOLOGIST
PHYSIOLOGIST	HEMATOLOGIST	OCEANOGRAPHER
CELL BIOLOGIST	MARINE BIOLOGIST	PODIATRIST
SCUBA DIVER	HERBALIST	ALLERGIST
OCULIST	CARDIOLOGIST	ANESTHESIOLOGIST
	OPHTHALMOLOGIST	

Name_____ Date _____

BOTANY AND ZOOLOGY

Life science, as you know, involves the study of living organisms. Botany and zoology are the two main fields of life science. Botany deals with plants, and zoology deals with animals.

Pretend you are a botanist or a zoologist. Choose a plant or an animal about which to collect information. In the chart below, fill in general information about your topic and any interesting details. Then write what other information you would like to learn about your topic. For more fun, do a full report on it and present it to the class.

Topic
General Information
1.
2.
3.
Interesting Details
1.
2.
3.
Other information I would like to learn about my topic
1.
2.
3.

Name _____ Date _____

PARTS OF A MICROSCOPE

A microscope magnifies small objects so that they can be seen easily. Label the parts of a microscope.

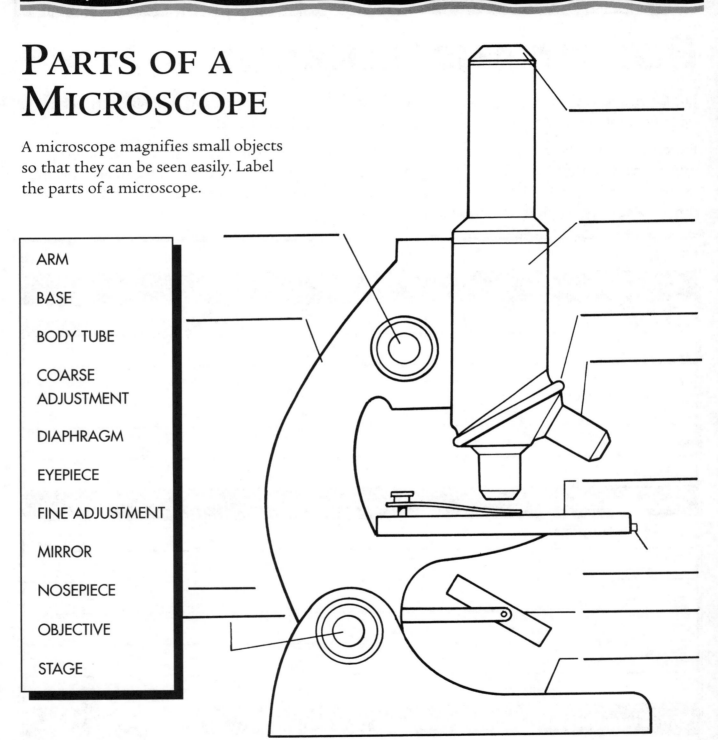

- ARM
- BASE
- BODY TUBE
- COARSE ADJUSTMENT
- DIAPHRAGM
- EYEPIECE
- FINE ADJUSTMENT
- MIRROR
- NOSEPIECE
- OBJECTIVE
- STAGE

List some things you would like to examine under a microscope. _____

_____

_____

Name_____ Date _____

EVERYDAY **life science**

Fingerprint Identifications

There are four basic types of fingerprint patterns: loops, whorls, arches, and accidentals. Three of the four patterns are pictured below. An accidental pattern has no specific form. Find out what kind of fingerprint pattern you have by completing the activity below.

LOOP

WHORL

ARCH

Materials Needed: a pencil, paper, transparent tape, a magnifying lens

1. Make a large smudge mark on the paper with a pencil.

2. Starting with your right thumb, rub your thumb on the smudge mark until it is coated with the pencil mark.

3. Tear off a small strip of transparent tape and press it onto your thumb.

4. Peel off the tape and stick it on another sheet of paper. Label this print "right thumb."

5. Repeat these steps with your other fingers. Label each print as you transfer the tape to the paper.

6. Using the magnifying lens, examine each fingerprint and identify the basic type of print. Complete the chart to the right.

FINGER	FINGERPRINT PATTERN
Right thumb	
Right forefinger	
Right middle finger	
Right ring finger	
Right little finger	
Left thumb	
Left forefinger	
Left middle finger	
Left ring finger	
Left little finger	

Was the pattern the same on all of your fingers? _____

Which pattern was dominant on your fingers? _____

Compare your prints with those of other class members by calculating the following:

_____ total number of loops in class   _____ total number of arches in class

_____ total number of whorls in class   _____ total number of accidentals in class

     FS-10616 Everyday Life Science

Answer Key

Page 1
1. C; **2.** K; **3.** D; **4.** J; **5.** I; **6.** M; **7.** N;
8. A; **9.** B; **10.** G; **11.** O; **12.** F; **13.** H;
14. E; **15.** L

Page 2
Answers will vary.

Page 3
Answers will vary.

Page 4
1. orangutan; **2.** chimpanzee;
3. manatee; **4.** chameleon; **5.** tortoise;
6. cottonmouth; **7.** porcupine;
8. rhinoceros; **9.** sandpiper;
10. blackbird; **11.** kangaroo;
12. groundhog; **13.** hummingbird;
14. hippopotamus; **15.** salamander

Page 5
1. C; **2.** S; **3.** A; **4.** G; **5.** B; **6.** P; **7.** E;
8. J; **9.** L; **10.** T; **11.** D; **12.** N; **13.** R;
14. F; **15.** I; **16.** K; **17.** M; **18.** Q; **19.** U;
20. H; **21.** O

Page 6
1. mountain; **2.** crowned; **3.** laughing;
4. bald; **5.** giant; **6.** hammerhead;
7. copperhead; **8.** rattle; **9.** prairie;
10. red; **11.** duckbill; **12.** flying;
13. monarch; **14.** star; **15.** jelly;
16. rainbow; **17.** sea; **18.** polar;
19. sand; **20.** grass; **21.** fire; **22.** army;
23. Earth; **24.** hermit

Page 7
KOALA; JAGUAR; PUMA;
MANATEE; GIANT PANDA;
VICUÑA; CHEETAH; COATI;
PRONGHORN; ORANGUTAN;
Answers will vary.

Page 8
Answers will vary.

Page 9
Answers will vary.

Page 10
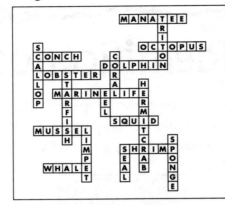

Page 11
Answers will vary.

Page 12
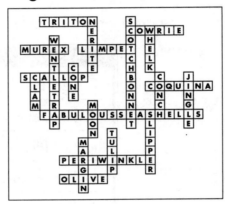

Page 13
Answers will vary.

Page 14
Answers will vary.

Page 15
Answers will vary.

Page 16
Answers will vary.

Page 17
1. RAVEN; **2.** ELEPHANT;
3. LEOPARD; **4.** ORANGUTAN;
5. VULTURE; **6.** BOBCAT;
7. TERRAPIN; **8.** CARIBOU;
9. CHEETAH; **10.** MONKEY;
11. OSTRICH; VERTEBRATES;
Animals with backbones or
spinal cords

Page 18
Answers will vary.

Page 19
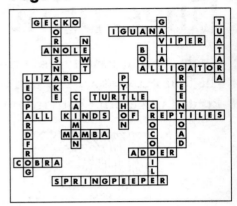

Page 20
Answers will vary.

Page 21
1. I; **2.** A; **3.** C; **4.** I; **5.** I; **6.** A; **7.** A; **8.** C;
9. C; **10.** I; **11.** I; **12.** I; **13.** C; **14.** C;
15. C; **16.** C; **17.** I; **18.** I; **19.** I; **20.** I;
21. I; **22.** I; **23.** I; **24.** I; **25.** I; **26.** A;
27. A; **28.** C; **29.** C; **30.** C; **31.** I; **32.** I;
33. I; **34.** I; **35.** A; **36.** A; **37.** I; **38.** I;
39. I; **40.** A; **41.** A; **42.** I; **43.** I; **44.** I;
45. A; **46.** C

Page 22
INSECTS—beetle, termite,
grasshopper, moth, bee, butterfly,
cricket, louse, ant, firefly, hornet, gnat,
wasp, fly, cicada, aphid, flea, mayfly

ARACHNIDS—tick, black widow,
scorpion, brown recluse, garden
spider, tarantula, mite

CRUSTACEANS—lobster, barnacle,
shrimp, crayfish, water flea, crab,
wood louse

Page 23
Answers will vary.

Page 24
Answers will vary.

Page 25
Answers will vary.

Page 26
Answers will vary.

Page 27
1. CARDINAL; 2. COLTS; 3. DOG;
4. SKUNK; 5. SNAKE; 6. FLEA;
7. TOAD; 8. MONKEY; 9. TIGER;
10. LION; 11. FOX; 12. SEAL;
13. LAMB; 14. BIRD; 15. COW

Page 28
1. TU; 2. G; 3. D; 4. G; 5. T; 6. G;
7. F; 8. TU; 9. D; 10. TU; 11. R; 12. G;
13. TU; 14. F; 15. R; 16. R; 17. TU;
18. T; 19. G; 20. D; 21. R; 22. F;
23. TU; 24. D; 25. F; 26. D; 27. R;
28. R; 29. TU; 30. F; 31. G; 32. D;
33. R; 34. G; 35. D; 36. D; 37. F;
38. G; 39. R; 40. R; 41. F; 42. TU;
43. T; 44. G; 45. F; 46. T

Page 29
PRODUCER—tomato, beet, carrot,
ivy, cabbage, asparagus, spinach,
potato, apricot, pecan, apple,
celery, grass, peach tree, bamboo,
seaweed, cactus, oak tree, moss,
wheat, rice, clover

CONSUMER—fox, spider, porpoise,
cow, squirrel, vulture, bee, wolf, pig,
manatee, earthworm, sheep, bobcat,
eagle, sunfish, turkey, deer, mole, wasp,
snail, elephant, elk, hummingbird

Page 30
Answers will vary.

Page 31
Answers will vary.

Page 32
Answers will vary.

Page 33
Answers will vary.

Page 34
Answers will vary.

Page 35
Answers will vary.

Page 36
1. BARREN; 2. BASKET; 3. BEAR;
4. BLACK; 5. BUR; 6. COW; 7. IRON;
8. LAUREL; 9. LIVE; 10. PIN;
11. POST; 12. RED; 13. WATER;
14. WHITE; 15. WILLOW;
16. YELLOW; 17. POISON

Page 37
1. AMERICAN HOLLY; 2. YAUPON;
3. SWEET BAY; 4. GREAT-
FLOWERED MAGNOLIA;
5. SHORTLEAF PINE;
6. LOBLOLLY PINE; 7. RED CEDAR;
8. COMMON JUNIPER

Page 38
Answers will vary.

Page 39
1. K; 2. G; 3. P; 4. L; 5. A; 6. E; 7. N;
8. B; 9. O; 10. C; 11. D; 12. I; 13. F;
14. J; 15. H; 16. M; 17. Q

Page 40
Answers will vary.

Page 41
Answers will vary.

Page 42
Most flowers have these four main
parts: calyx, corolla, stamens, pistils.
The parts and elements of the
students' chosen flowers should
be labeled.

Page 43
Answers will vary.

Page 44
1. b; 2. a; 3. c; 4. a; 5. c; PLANTS
ARE AN EXCELLENT SOURCE OF
MINERALS AND VITAMINS FOR
A HEALTHY BODY.

Page 45
Answers will vary.

Page 46
Answers will vary.

Page 47
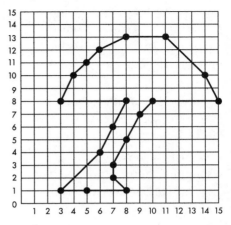

mushroom

Page 48
Answers will vary.

Page 49
Answers will vary.

Page 50
Answers will vary.

Page 51
Answers will vary.

Page 52
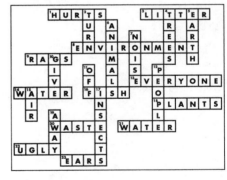

Page 53
Answers will vary.

Page 54
1. CARBON MONOXIDE, A;
2. PARTICULATES, E;
3. NITROGEN DIOXIDE, C;
4. HYDROCARBONS, D;
5. PHOTOCHEMICAL SMOG, F;
6. SULFUR DIOXIDE, B

Page 55
Answers will vary.

Page 56

organs—heart, gall bladder, intestines, esophagus, lungs, liver, stomach; endocrine system—adrenal gland, pituitary, thymus, parathyroid, ovaries, thyroid, testes; diseases—hepatitis, influenza, smallpox, measles, colds, tetanus, tuberculosis, rabies; body functions—circulation, digestion, nerve reactions, respiration, reproduction, excretion, movement; skeletal system—arms, legs, neck, collar bones, spinal column, pelvis, shoulder blades, sternum, skull, hands, feet

Page 57

1. MOUTH; **2.** ESOPHAGUS; **3.** STOMACH; **4.** SMALL INTESTINE; **5.** LARGE INTESTINE; **6.** RECTUM

Page 58

Answers will vary.

Page 59

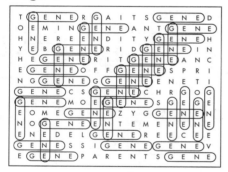

1. TRAITS; **2.** DOMINANT; **3.** HEREDITY; **4.** HYBRID; **5.** INHERITANCE; **6.** OFFSPRING; **7.** GENETICS; **8.** CHROMOSOME; **9.** ZYGOTE; **10.** MENDEL; **11.** RECESSIVE; **12.** PARENTS

Page 60

2; 7; 6
9; 5; 1
4; 3; 8
15

Page 61

iron—Fe; fish, meat, raisins, eggs, green leafy vegetables; carries oxygen to body cells in hemoglobin

calcium—Ca; milk, cheese, green vegetables; bones, teeth, blood clotting, muscles, nerves

iodine—I; iodized salt, seafood, fish; required by the thyroid gland in thyroid hormone

phosphorus—P; meats, whole-grain cereals, milk, egg yolks, fish, peas; bones, teeth, muscle and nerve activity

potassium—K; bananas, fruit, vegetables, cereals, meat; muscle contraction, transmission of nerve impulses

sodium—Na; table salt, butter, bacon, vegetables; regulation of body fluids, blood pressure

magnesium—Mg; nuts, yeast, cocoa, whole-grain cereals, beets, blueberries; muscle and nerve activity, bones, enzyme function

manganese—Mn; nuts, peas, oat flakes, wheat germ, coconut; growth of bones, enzyme activator

copper—Cu; liver, kidney, nuts, raisins; maintenance of bone, nerves, and connective tissue

sulfur—S; dried fruits, nuts, eggs, beans, cheese, grains, cabbage, onions; hair, nails, skin

Page 62

Answers will vary.

Page 63

Answers will vary.

Page 64

Answers will vary.

Page 65

1. B; **2.** B; **3.** B; **4.** B; **5.** V; **6.** B; **7.** B; **8.** V; **9.** V; **10.** V; **11.** V; **12.** V; **13.** B or V; **14.** V; **15.** V; **16.** B; **17.** B; **18.** B;

19. B; **20.** B; **21.** B; **22.** V; **23.** V; **24.** V; **25.** V; **26.** V; **27.** V; **28.** B; medications, thorough washing of food and hands, disinfectants, use of masks, vaccinations; vaccinations, avoiding contact, drinking clean water, avoiding animal bites

Page 66

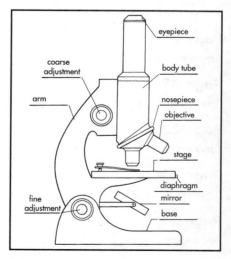

Page 67

1. INFLUENZA; **2.** CHICKENPOX; **3.** LYMES DISEASE; **4.** PSORIASIS; **5.** EMPHYSEMA; **6.** DIPHTHERIA; **7.** YELLOW FEVER; **8.** TUBERCULOSIS; **9.** PNEUMONIA; **10.** PARROT FEVER; **11.** TYPHOID FEVER; **12.** BUBONIC PLAGUE

Pages 68–74

Answers will vary.

Page 75

eyepiece
coarse adjustment
body tube
arm
nosepiece
objective
stage
diaphragm
fine adjustment
mirror
base

Page 76

Answers will vary.